SUPREME TECHNOLOGY: HIDDEN ADVANCEMENTS BLACK BUDGET WITHIN

BY DUANE WHYTE

Quantum leap

A huge often sudden increase or advance in something.

Table of contents

Dedication page	Pg. 7
Preface	Pg. 8
Chapter one: Cutting Edge	Pg. 10
Chapter two: Innovative entertainment	Pg. 15
Chapter three: High-tech Gadgets	Pg. 22
Chapter four: Medical advancements unbelievable breakthrough	Pg. 27
Chapter 5: Sophisticated military Weapons	Pg. 31
Chapter 6: New children's creations	Pg. 35
Chapter 7: Brand new articles worth having.	Pg. 41
Chapter 8: Schools and educational leap	Pg. 47
Chapter 9: Back II the Future	Pg. 52
Chapter 10: Exploration	Pg. 58
Closure	Pg. 61

Dedication

This book is for open minded individuals who believe in more than what meets the eye and the possibilities that exist in the world and beyond the stars.

Preface

If you are reading this book right now, I'm going to enlighten you with some amazing information about supreme technology that is deliberately hidden from the general public. When it comes to keeping classified secrets from humanity our shadow government operates and thrives in their secrecy. This is how they maintain ultimate control and power over the planet. It is actually a disservice to the population of people to not disclose the truth about hidden supreme technology that has physically already been manifested.

It is very unfortunate we live in a world of have and have nots. The one percent ruling elite class sits at the top of the pyramid and dictates world affairs and share all their secrets among each other about Quantum leap supreme technology that could be astronomically beneficial for humanity. The quality of life for the masses of people would increase dramatically. It would change human being lives forever.

How many have you heard of black budget? It is a secret classified government operation facility located somewhere sixty feet underground in Roswell, New Mexico. This black budget program has unlimited capital of 30 billion dollars a year funded by our government. This black budget program is a big part of the shadow government when it comes to protecting hidden supreme advancements in technology. Believe it or not black budget has already developed and designed new innovative technology beyond the general public comprehension to even fathom what is really out there.

This top-secret black budget program has more than 1,600 German scientist engineers, technicians, mathematicians, and architects. They all come together to consolidate and brainstorm on creating the most incredible technological advances unknown to man. This state-of-the-art facility also works alongside higher overseers who assist in these Quantum leap advancements not available to the masses of people on the planet. As we indulge into these secrets developed hidden supreme technological advancements that has already been manifested. Readers understand our government is good at hiding the truth and only reveal advancements in small increments at a time.

Just know in actuality the black budget program is 365 years advanced when it comes to supreme hidden technology.

Chapter One: Cutting Edge

1. We all have experienced constipation trying to move our bowels in the bathroom before it can be an extremely uncomfortable feeling being backed up and cannot release yourself. With all the remedies and medicine available sometimes these products do not work or take effect in a timely matter that you need it to. Black budget has created a spray called Commode Go this works by spraying a chemical agent that releases phosphate particle ions allowing you to soften the stool and moving your bowels.

2. Taking a shower is an everyday thing a lot of people do that typically includes some type of hand soaps or body wash. This would be known as the common way of taking a shower. To your surprise black budget has developed a shower head with organic lather cleaning soap called Compound Suds. This ten-inch shower head has a source filter that combines the water

with the cleaning agent allowing you to wash yourself in a new experience way.

3. Coin Counter Piggy Bank sorts and organize your change into the right dispensary slot. Black budget took the same concept and developed the Med Bank. It is similar to a coin counter piggy bank except the Med bank organize and sort prescription pills this works by typing in the meds and the dosage you are taking on the digital med bank device that has a built-in smart memory chip. This Med bank also comes with a timer to alarm you that it's time to take your medication.

4. Pens are used every day for writing some people even collect them as novelties. We are long way from fountain pens. Black budget has developed and designed a pen called the John Handcock memory pen. This one-of-a-kind unique pen has a recorder on it when you speak into it the pen device it allows up to twenty characters words at a time to be recorded. All you have to do is hold the pen to the paper and watch the magic of the pen device starts writing down what you said. This pen could be a huge benefit for kids learning how to write,

people with arthritis or maybe even the blind.

5. Women who go to the hair salon knows a good thing or two about sitting under a hair dryer for 15 to 30 min just to have completely dry hair. Women would love a much faster way to dry their hair. Black budget has developed and created a quick hair dry spray bottle called conditioner quick dry. This ten-ounce ten plated aluminum canister is fused with micro particles that spins up to a 190 degrees Fahrenheit instantly drying your wet hair within seconds.
6. Chairs are used every day for many different reasons. It could be used for eating, working, or just relaxing. Some chairs are more comfortable than others depend on the design and style. There is actually a whole new way of experiencing comfortable relaxation while sitting down. That's because black budget has developed an innovated design chair called water bench. This one of kind unique chair is made from silicone, latex and has oxygen water capsules inside the sealed tight cloth material chair allowing you to enjoy sitting down in the most exhilarating way.

7. A mirror is a glass that can show your image reflection. Mirrors are used for many different reasons. Mainly to see ourselves and to make sure we are presentable before leaving the house. Just know we're a long way from the conventional way on how we use mirrors. Black budget has designed and developed an extraordinary interacting looking glass called Mirror Mirror on the wall. This unique hologram laser beam is implanted in the looking glass compound with AI dimensional intelligence. When you turn this customized AI looking mirror it can greet you with confident uplifting words of the day such as you are so beautiful," "You know your sexy right," "You're going to have an amazing day." Those are just a few examples with this preprograming AI looking mirror can do. There's also a setting button on the looking glass that allows you to set your hologram image to male or female host. You also have the option to preset all the things you love about yourself with this predesigned mirror.

8. Putting money in a banking institution is a secure safe way to protect your economic funding. We use banks to withdraw or deposit money. Sometimes getting to the bank can be an inconvenience if you're having car trouble so making a trip to the bank is not an option. Those days could be over immediately; Black budget has developed a portable cash receiver machine called fax me funds. This unique incredible technology allows you to receive cash right from the comfort of your own home. This works by using the home fax me fund machine that is connected to your bank that has your routing and account number. All you must do is digitally type in your name and account trying to withdraw or deposit and this internal system software will print you real money in minutes.

9. Millions of people play tennis a year. People who play by themselves need a portable ball machine injector to engage back and forth. Black budget has revolutionized a single person can play tennis now that is because they have developed and created the author ash hand rack post encounter machine. This high-tech robot has a custom built in forty-point fourteen Tesla magnet inside the robot allowing you to hit the ball and receive it right back. This interacting machine robot

will feel like a real life one on one interacting tennis match.

10. Bike riding is a great way to get some exercise. There are many health benefits when it comes to riding a bicycle. Sometimes riding a bike for an awfully extended period of time can make you too tired to pedal uphill or just at all. Black budget has invented lazy pedals. This new dynamic bike design allows riders to pedal freely without physically moving their legs. Lazy pedals will do everything for you giving you a whole new experience how you ride bicycles.

Chapter two: Innovative entertainment

1. In 1997, There was a game console called the Atari 2600. This was the first gaming system that allowed you to insert a cartridge game and play. We are a long way from the traditional way we entertain ourselves with gaming systems. Black budget has developed a gaming pod simulator. This walking pod simulator allows you to experience gaming on a whole new level. It feels like your literally inside the game. This pod is equipped with a smart chip built inside the helmet that connects to an

electrical charge controller suit that you must put on. This can be an amusement of fun for countless hours.
2. Watching television can be amusing and entertaining to the consumer viewer. We have all watched tv commercial advertisements and wished we could instantly order products and food directly off the tv screen. People this might sound so far fetch to even physically possible to even exist. Black budget has materialized and developed the smart portal shop vision. This one-of-a-kind unique high-tech portal tv can have a pipeline that tunnels from commercial establishments right to the comfort of your home. Owning this incredible smart portable shop vision is not cheap because installation must be done. This system cost a hundred grand to set up in order to directly place orders from tv to home.
3. Everyone loves to dance it's a good way to burn calories and enjoy yourself on the dance floor. Unfortunately, not everyone has rhythm. So, people who do not have the natural ability to dance on beat is usually off key. This could actually be the thing of the past because black budget has designed and created robotic rhythm shoes. These incredible

one size fit all unique shoes will teach you rhythm in step patterns allowing you to learn how to dance to the rhythm of the beat. These robotic rhythm shoes are implanted with synthetic Mach five AI technology that will control your balance making it easy to learn how to dance. You may even be the life of the party on the dance floor.

4. When we think of the idea of traveling, we usually consider trains, planes, and automobiles. These are typically the three ways the majority of American people choose to commute. Unfortunately, this could be an incredibly stressful time trying to gather yourself and family to pursue the journey of traveling around the holidays. Because everyone has the same mindset so patience is the one thing you will need to deal with the holiday traffic on your way to your destination. The congestion of dealing with a lot of people could become nonexistence. Black budget has created a whole new way of traveling with the invention of the Teleportation zap. This innovative incredible machine pod allows you to arrive at your destination in seconds. It's an exciting new way of traveling designed to break your molecules down

to a single atom allowing your stardust flesh to transmit your particles instantly. This Syfy technology has been in the world for general population use since 2010 patents are still pending.
5. How many of you remember playing a physical activity game called twister? This fun home game will stretch your body in so many different ways. This twister home game usually would require other bodies to participate. Black budget has changed that with an updated version called Twist me on 2.0, Now without any actual people you can still have countless hours of fun by yourself. All you have to do is turn on this Mach three electrical robot mat and watch it come to life. It has setting option buttons so you can control how many hologram players you want to join in. You have up to four interacting hologram players to choose from.
6. A racetrack toy set for kids was invented in 1965. This racetrack toy set came with two joystick, two cars and electrical fuse tracks that you use to snap together also an adapter to plug into the wall. Once you would put the car tracks together than you would plug up all the connectors to the racetrack such as joystick than you would plug the adapter

into the wall circuit. Next you put the cars on the electrical charge track then you would control the speed of the car with electrical joysticks. For kids this would be countless hours of fun. Black budget has developed a new version way for playing with this toy racetrack for kids with the invention of the speed way head band. This new innovative speed way head band doesn't require any connectors or wires. You actually will be controlling the cars on the track with your mind using the electrical telekinesis head band.

7. Jogging has many health benefits. It's an effective way to lose weight, stay in shape and build strong bone muscle. Black budget has developed the runaway shoes that allows you to run for hours without getting tired. These incredible one of kind unique shoes have solar power energy in soles panels inside them that is charged by the sun. These shoes generate electricity and activates when you start running. This new way experience of jogging will seem effortless to your body. You will almost feel not human with these high-tech incredible runaway shoes.

8. Frisbees is a fun hand throwing plastic disc you can toss back and forth with

someone outside. This can be a lot of fun with multiple people. Black budget has designed a new way with playing with frisbees even by yourself if no one can accompany you. They have developed the Frisbee back. This work by putting on a magnetic wrist band around your wrist. This magnetic wrist band is connected to the frisbee that has a pinpoint laser chip inside allowing you to throw this frisbee 800 yards away and directly comeback to you just like a boomerang. Because of the laser identifying chip inside the wristband, it will always pinpoint your target with accuracy.
9. Combing and Brushing hair is something a lot of people always do. It activates sebaceous gland that produces your scalp natural oils keeping your hair naturally moist. It's a good way to maintain a healthy scalp. Black budget has invented the talking hair counter comb. Every single time you comb your hair this digital talking comb will count out how many times you groomed your hair. This also comes in brush form designed for men who like brushing their hair to achieve ways. This comb and brush have a sensor smart chip voice recorder implanted inside these unique

one of kind comb and brush comes with a charger. This would be a new and fun way to enjoy combing or brushing your hair.
10. A pull up is an upper body strength training exercise. In order to do this exercise activity, you must hang on to a bar with both of your palms. This takes a lot of upper body strength to do. A lot of people find this exercise very difficult. Black budget has designed the pull up bar gloves. These incredible gloves will allow you to lift your own body weight effortlessly making you feel stronger than you ever have before. These pull up bars' gloves are fused with a robotic skeleton frame that has sensors inside that will balance your body. The fuse mechanism inside these high-tech gloves is extraordinary giving anybody the ability to do pull ups.

Chapter three: High-tech Gadgets

1. Snorkeling is the practice of swimming underwater. A lot of people like to do

especially if you want to swim with exotic fishes. To achieve this fun underwater activity, you must be equipped with a snorkel mask or a shaped breathing tube. This conventional way could be a thing of the path. Black budget has developed a one of kind tablet called the smirched oxygen pill. This extraordinary pill must be swallowed 10 minutes before you dive and snorkel under water. Once the pill is activated it will release oxygen bubbles in your lungs allowing you to breathe underwater without any mask or tubes. You have up to half an hour oxygen time before your submerged oxygen pill runs out.

2. A washer and dryer for clothes is essential when it comes to cleaning clothes. Sometimes when we dry our clothes, they may seem dry but it's not that's when you have to start the spin cycle again to get your clothes fully dried. Black budget has developed a revolutionary version way called the iron spin 2000. This high-tech innovated technology allows your dryer machine to spin 200 billion times per second completely drying your clothes in less than five minutes no matter how small or big your load is. This product has been developed for over 20 years already and

could be on the market today if they allowed it.

3. We have all used a hammer to nail something together before one time or another. I'm sure we have accidently hit our thumb with the hammer, and it really can be a painful experience. This could be obsolete black budget has invented and developed and awesome hammer glove. This one size fit all hammer glove is designed with lace stainless steel metal particles on the right-hand knuckles and once you put on this hammer glove you will be able to nail anything together without ever hurting yourself again and feel extremely powerful with the hammer glove on.

4. A hand ball is used for many different fun outside activities. You can throw this ball back in forth with someone. You can bounce this ball against the wall and play by yourself. There are so many different ways you can have fun with the hand ball. Black budget took the same concept and developed and created the Bolloy me ball. This Bolloy me ball can levitate eight feet from the ground and follow you wherever you go. It can also interact with you if you hit the ball, it will come three feet right back to you. This Bolloy me ball has jet motion sensors

inside with an electrical magnet conductor that generates powerful levitating energy. The Bolloy me ball also has a camera on it so you can take selfies and ten-hour battery life charge. This one of kind ball will make you feel like you always have companionship around you.

5. Everyone has doors in their homes that we walk in and out of all the time. Sometimes trying to carry items and open the door can be a challenging task to manage. That's why black budget has developed and invented the master door sensor. The double plated custom doors allow you to walk in and out of doors hands free. The doors have facial and voice recognition and accepts voice commands within a six feet radius.

6. An oven can serve many different purposes; One maybe to cook, bake or just keep your food warm for whenever you are ready to eat. Imagine a refrigerator and oven combine together. Black budget has designed the ovenator. This smart ovenator chamber is so high-tech it can store food hot or cold at the same time. It has a food profile setting button so you can identify your grocery items. And when you're ready to eat or cook a meal this ovenator chamber can do all these things and more. This will be a

game changer once it's available to the general public.

7. A jacuzzi is one of the most relaxing experiences if you ever been in one. It comes with jet temperature adjustment buttons you can control. Black budget has revolutionized the jacuzzi experience with water hands. This hot tub takes enjoying and relaxing in a jacuzzi to whole new level. That's because water hands have eight real life feeling hands made out of synthetic AI. There are four synthetic water hands one each side of the jacuzzi feeling like you're getting a full complete body massage by chiropractor under water.

8. Hearing aids are a great device to help people hear better when they experience hearing loss. This tiny ear device can maximize and enhance your hearing capabilities by amplifying the sounds around you. Black budget is light years away with the creation and development of owl hearing aids. This small android microchip will give you incredible hearing ability to hear 100 feet away. It will seem like you have super hearing powers.

9. A hologram concert gives a fan a chance to still interact with their favorite musician that

has passed away. The hologram concert also gives the deceased artist a chance to be on stage again to perform their hit songs. Bringing a deceased musician back to life on actual concert stage can cost at least a 100 thousand dollars. Black budget has developed and created that at home concert box preprogrammed with your favorite artists who can come to life straight out the concert box and perform their hit songs. This works by turning off the lights in your room and then turn on the three-dimensional hologram concert box and watch your favorite musician come to life instantly and perform for you.

10. When we think about a genie in a bottle, we think about three wishes that are granted to you by mythological spirit. Black budget has taken the same concept and developed the talking genie in the bottle. This voice controlled virtual assistance is a similar idea to Alexa accept this talking genie in the bottle releases smoke and can light up ten different colors. This device can play music and interact with you and can be entertainment for everyone.

Chapter four: Medical advancements unbelievable breakthrough

1. Normally when we get sick, we go to the doctor to see what's wrong this can be time consuming. If you work or just too busy to go to doctor for a checkup about your help. Convenience means everything to people who are too busy in everyday life. Black budget has created a high-tech extraordinary pill called Med scanner. Med scanner is a pill that has a smart chip and camera inside that can diagnose any internal issues you may be experiencing. Once you swallow this Med scanner pill it takes up to twenty-four hours to pass through your system giving your internal body a complete analysis and uploaded to your computer and sent to your primary doctor to look at your results.

2. When it comes to blindness it can occur suddenly over a period of time due to health issues. Being unable to see can be challenge for anyone that lost their eyesight. Assistance is necessary whether you use a human helper or seeing eye dog. Black budget can eliminate those few options with development and design of the spec vision contact lens. These futuristic dynamic contact lens has bionic technology and a smart chip implanted in these contact lens. To achieve eyesight vision again you must have these spec vision contact lens surgically attached to the cornea of your eye

giving you the ability to have your independence again.

3. Being paralyzed can be a real unfortunate situation a lot of times it's due to an awfully bad car accident. People who experience losing their ability to walk may become depressed due to the fact they now have limited mobility and have to rely on a wheelchair to get around. Black budget has created and developed the Bionic spine. This Titanium metal muscular skeleton spinal cord bionic suit is connected to your vertebrate surgically giving you the ability to walk, run or jump again.

4. Organ donations saves lives. One deceased organ donor can save up to eight lives. Unfortunately, eight thousand people die a year because they're not able to get an organ transplant in time. Black budget has created and designed the five D human organ maker. This incredible human organ tissue maker machine can print and duplicate any organ in the body using stem cells.

5. Aging is a part of growing old. It's a natural process every living human will experience. We all wish there was a way to prolong our beauty with the fountain of youth. Black budget has created the forever skin on. This

chemical is a synthetic imitation of natural looking skin. You apply it like lotion giving your skin a non-aging look for months. This human synthetic skin lotion want wash away in the shower either. It's waterproof.

6. Condoms were in invented in 1855 by Charles Goodyear. The whole purpose for inventing condoms was to protect men from STD's and unwanted children. Unfortunately, only thirty percent of men wear condoms. Black budget has developed protection for men called prophylactic spray. This one of kind unique spray can coat your penis with a condom that only takes one minute to dry after you spray it on.

7. Having children is a gift from God. It's the most incredible experience for a mother and father to witness. Unfortunately, some couples have difficulty achieving this goal naturally. There are other means to having children some do intravenous, adopt and some even have surrogate. There's a whole new revolutionary way to have kids. Black budget has created and invented the baby pod. This egg-shaped machine can literally grow a real-life baby within nine months. Just like a plant you have to water it; keep it temperature controlled this baby pod replicates a woman womb.

8. Medical beds are used for patients who are admitted into the hospital. These beds typically have control buttons you can use adjust your comfort level or even call for help in case of emergency. Some patients who are bed ridden have a tough time getting up going to the bathroom themselves and may need some assistance. Black budget has created and designed PBT (The patient bed toilet). This state of the art one of kind bed allows patients who are bed ridden the ability to use the bathroom on their own. The PBT has a commode pipe attached to it that can flush urine and feces. The PBT has self-automatic sanitary spray that cleans you after use.

9. Getting shot is a painful experience if you ever been through that traumatizing situation. Sometimes when people get shot bullets can travel through the body and can be lounge in area of the body making it hard to remove. So, a lot of times the doctor just leaves them in to avoid further damage to the internal body. Black budget has invented the bullet eradicate magnet pen. These hundred forty-millimeter pen magnets have the power of thirty-three-point fifteen Tesla electrical charge technology that can

penetrate any foreign objects out of the body that is hard take out manually.

10. Dialysis is for people who have renal failure and need to get their blood cleansed. Needles and bleeding are part of the process when comes to getting on and off the machine and stopping your bleeding can take some time. Black budget has created the mini ball glue sponge. This one and half inch mini ball glue sponge can instantly stop your access bleeding once put on properly.

Chapter 5: Sophisticated military weapons

1. A grenade is a small bomb thrown. This weapon is designed to kill on impact with a powerful explosion if you're within range. Black budget took the same concept and created and developed the LNB (Liquid nitrogen bomb). This 2.5 diameter steel case sphere is packed with liquid nitrogen and once you release the lever and throw it. It will instantly freeze the enemy right where they stand.

2. Mice has been around for millions of years. These tiny rodents don't have bones so they can get in tiny holes and squeeze through any gap. Black budget has weaponized these rodents. They are called mice control. These mice have cameras and microchipped bombs inside of them sending them inside the most remote locations that are hard to get to humanly possible. Deploying these mice can destroy enemy and his headquarters.
3. When you think about a water gun you associate that with toys and outside summer fun activity. Military weapons are really advance black budget has designed and developed the acid shooter. This incredible dangerous gun can liquify the skin on contact with a chemical fluor antimonic acid leaving any part of your body hit permanently damaged.
4. Parachutes are a big part of the military use when it comes to surprise attacks on the enemy. It also deploys soldiers safely to the ground. Black budget has revolutionized the parachute with the camouflage pack. This extraordinary parachute is made out of quantum stealth technology allows you to be invisible once deployed in the air.
5. Soldiers who are in the military risk their lives every day to defend Americans. casualties or getting shot is a part of the job

description when fighting on the battlefield. Usually, when a soldier is wounded from the gunfire you must pack guards directly to the wound to stop the bleeding. Black budget has designed and developed the trauma stop bandage. This unique one of kind bandage stops bleeding instantly up to five hours. This trauma stop bandage has AI technology fused in it with sterilization ointment. Once you place it on the wound its acts as a self-guiding medical aide sealing your injury from further damage.

6. Being a soldier in the military a M16 is the standard use when it comes to combat. When you're at war with the enemy the last thing you want to worry about is running out of ammunition. This could mean life or death for an army regiment. Black budget has invented the 3D portable bullet maker. This incredible 3D portable bullet maker can duplicate a hundred 3D bullets in a minute for your weapon or any other artillery gun you may own. So, you're always ready on the battlefield.

7. The military thrives on cutting edge technology when it comes to weapons such as tanks, machine guns and missiles. Military weapons are designed for mass

destruction and casualties. Black budget has developed an extremely dangerous chemical gas agent weapon called dose off. This 9,007-hundred-pound bomb can put people in permanent sleep within a sixty-block radius. Releasing this chemical dose off bomb is painless but deadly.

8. A military helmet is used to protect a soldier's head. It serves as a piece of personal armour during combat. It's durable, bullet proof and designed to stop brain damage that was the main purpose use for a conventional military helmet. Black budget has upgraded the military helmet with new equipped AI technology called SHP helmet (The survival head protector). This incredible helmet has eight cameras with motion sensor bomb detection that can help you pinpoint the enemy hiding in blind spots you cannot see. This helmet can increase your survival rate.

9. Military Binoculars are used for various reasons such as surveillance, combat or hunting. These field glasses allow you to see distant objects. This telescope has been weaponized. Black budget has turned the M22 military binoculars into a sophisticated sniper lens called dead eye. This incredible dead eye weapon comes with four

preloaded standard M80 ball round bullets. These binoculars can shoot up to 300 miles away. This dead eye weapon is very accurate when sites on target at great distance.

10. Army stretchers are primarily used in the military for search and rescue or those mortally wounded in battle. It takes manpower to carry a soldier. Black budget has developed a new version stretcher called avatar bed. This awesome metal and aluminum avatar bed is an artificial intelligent robot with a built-in smart chip that allows itself guiding stretcher to carry injured or deceased soldiers back to base all on its own.

Chapter 6: New children's creations

1. A baby first steps can happen within a few days or a few months. There are many products on the market that can help teach babies how to walk. Nothing compares to the new invention design. Black budget has created a pole device called the baby self-balancing walker. This one-of-a-kind baby pole walker helps teach babies how to walk. This self-balancing pole has a motion sensor

frame with electrical magnets inside no assistance is needed. All you have to do is turn it on and put baby hands on the pole. This self-balancing walker will teach than to walk earlier than they should.
2. Early child brain development begins with learning educational tools that can teach babies the basic fundamental things. There are many learning devices on the market that can assist your child in early learning. Black budget has a game changer with the design and invention called baby learn mic check. This portable AI android mic has a built-in memory database with voice activation record playback. There's also a setting mode that teaches babies their ABC's, name, home address and animal sounds. This baby learn mic check will increase your baby's brain by 80 percent within 21 weeks old.
3. A diaper is what babies wear before they learned to become poddy trained. It's a piece of absorbed material wrapped around a baby bottom. Parents with infants know all too well about the stanky task that has to be done to keep their baby clean. Black budget has developed a new baby bottom wrapped material called the baby diaper self-changing and cleaning machine. This bowl-shaped machine is similar to a toilet if your baby does a number one or two all you have to do is put your baby in a self-cleaning and

change bowl preloaded with a fresh diaper and add warm water and soap. This smart baby cleaning device will sanitize your baby in five minutes eliminating the stanky task.
4. Baby cries to communicate their either hungry, bored or experiencing discomfort. Sometimes it's hard to understand what babies want when they cry for attention. Black budget has designed a bracelet called calm down. This smart chipped bracelet has electrical emotional sensors inside that can measure your baby cries and can identify whether your baby wants to eat or diaper change or not feeling well. This calm band will eliminate any issues your infant might have.
5. Sudden infant death referred to as crib death can be associate with low breathing oxygen when baby sleep face down in their crib. This can be very unfortunate horrible situation for parents who have lost their infants. Black budget has developed a new invention called the safeguard carriage. The safeguard carriage has a pressure motion rotating electrons embedded inside that can detect whether your baby is distressed or uncomfortable and any problems they may be experiencing while sleeping. The safeguard will protect them from danger.

6. Baby brains responds to faces that's why they love to stare at you. It really gets their attention just watching your facial expressions. The way you move your eyebrows or makes sounds with your mouth. Infants learn language signal emotions when your happy, sad or mad. This is the most important social stimulant for a baby. Black budget has designed and invented the baby mimic robot facial. This incredible robot baby mimic facial can interact with an infant in real time with tendency to imitate communication directly assisting an early child brain development.

7. When it comes to keeping babies entertained there are a lot of products on the market, such as rattle and sensory teether or a baby ring stacker. These are just a few playing gadgets that can keep your baby amused or distracted for a certain period of time. Black budget really has the game changer with the invention of the crayon lights baby shades. These smart shades are fused with 366 electrical crayon colors. When a baby put these shades on, they will experience a multitude of pigmentation that will keep them excelerated for hours.

8. A woman body undergoes many different transformations during her nine months of

pregnancy and protecting the development of an unborn child is a mother's number one priority. That's why sonograms are important to have so it can detect any prebaby problems your baby may be experiencing. This actually a whole new revolutionary way to detect any issues with a woman pregnancy. Black budget has created the mommy womb monitor patch. This artificial intelligence adhesive skin patch can detect any harmful stress levels or issues your unborn baby might be experiencing. This works by placing the mommy womb monitor patch on a pregnant woman stomach. This monitor patch changes colors if any precondition starts to mature. Blue means unborn baby is hungry. When the patch turns pink preterm can be starting. When it turns red early defects can be manifesting. Mommy womb monitor patch can help reverse any preconditioned problems during a woman's pregnancy.

9. As parents having kids is responsibility that can bring many challenges and the just thought of not being there for your child or children can be very devastating. A loss of a parent or parents can be a tragedy in an unfortunate situation not being around physically to raise and watch your kids grow up to be adults. Whether it's from health

issues or a bad car accident. Children who lose their parents at an early young age can experience psychological trauma knowing they have grown up alone in a world without a mom or dad. Black budget has invented the I will always love your memory recording phone called Titanic. This one of kind memory recording phone is an insurance policy for parents just in case something bad happens to them. This voice recognition recording memory phone has a mom and dad message to you button. You can prerecord up to twenty minutes of your voice to say how you feel about your child and if their listening to parent's voice recording that means they're not physically here on this planet anymore. You just want your child to know your sorry, but you will always love them with the instrumental in the back playing. This Titanic I will always love you memory phone will be a great investment.

10. Double Dutch is a jump rope game played with two long jump rope that swing in opposite directions. Two kids must hold the rope on each side so another person can jump in the middle skip and jump and try not to get caught up in the ropes that moving fast. Timing and having rhythm is the object to this game. Black budget has revolutionized the double dutch game with a

new version called AM double D boxes. These 14 feet ropes are self-rotating motion sensor rope boxes can automatically turn these jumping all on it's on no manual help needed.

Chapter 7: Brand new articles worth having.

1. A baseball is used in sports of baseball to hit the ball after its thrown by a pitcher. The umpire is usually the one who calls the strike outs if you don't have one it's ok. Black budget has invented an advanced electrical AI talking metal bat called Ratchet chat bat. This one of kind Ratchet chat bat can say strike, homerun and foul ball every time a ball is thrown, and you swing. This incredible ratchet chat bat can also determine how fast the ball was thrown to you and more.
2. As humans we go to the toilet to eliminate waste from our bodies. This is a natural recurrent thing we must do to clean out our internal body system. There are a lot of different circumstances where someone may need assistance going to the bathroom to do a number two. Such as someone with debilitating old age or someone who doesn't have no arms or people with real bad arthritis. Black budget has invented the automatic buttocks wipe toilet called finish me. This incredible custom design toilet has

a self-guiding AI synthetic robotic hand that can mimic a real human hand function that is connected to the floor flamage inside the toilet for assistance. All the person has to do is press the finish me button and then AI robot hand will thoroughly clean your buttocks and all the person has to do is flush the toilet and stand up when their done.

3. When we think about a vending machine we usually thinking about dispensing snacks, beverages or even cigarettes. The main purpose of a vending machine is to hold items inside. Black budget took the design vending machine further with the creation of the home furnish gem maker. This state-of-the-art custom jewelry has a built-in wax 3D printer with engraving software settings. All you have to do is program the material you want to use from metal, gold and silver hit the hydraulic press button and watch the magic of this furnished gem maker customize your jewelry.

4. In cold weather people usually wear coats when going outside. Winter coats keep you warm because of the material inside. Imagine going outside during wintertime without a warm garment. Black budget has created and developed an advance coatless body cream called polar bear. This

incredible skin agent cream can keep you warm without a coat up to 24 hours. This polar bear cream can insulate and retain your body heat with a ten-centimeter artificial blubber.

5. The primary reason why woman wear makeup is to enhance their appearance for a better complexion tone. This can be done by yourself manually or you can go to a professional makeup artist for the perfect face coding. Black budget has invented a makeup mask applicator called make up express. This one of kind make up applicator can transform your looks in minutes. This works by putting on foundation express mask which has a hundred different color settings. All you have to do is pick the face color settings you desire and watch the magic of the foundation express give you the perfect application in return.

6. There's nothing like buying a fresh pair of sneakers. It can be a great investment if you can maintain that fresh brand-new look. Weather you have one pair of sneakers or many pairs of sneakers. People who take pride in their shoes want to always keep them looking fresh at all times. There's a lot of products on the market to clean your sneakers and maintain their quality even if

they may have been broken in. Black budget has invented the automatic sneaker steam cleaner called refresh kick box. These eleven points seventy-five length box has a built-in vapor pressure rise steam heater with a brush sanitizer attachment. All you have to do is plug this plastic sneaker container on place your pair of shoes inside and add water and a magical cleaning agent inside the nozzle holes and within ten minutes your sneakers will look as brand new as if you just purchased them.

7. Tying a tie is not the most simplest thing to do for a lot of guys. There are a lot of video seminars that can assist you and teach you have to do this but sometimes it still doesn't help. Black budget can eliminate all of this confusion with the design and invention of the automatic tie wire called connection bind. This incredible one of kind memory smart chip wire has a recognition magnet mechanism inside. All you have to do is place the memory smart chip wire inside of the shell interlinings push it all the way through then place the tie around the shirt collar. Push the button on the self-guiding smart wire and watch the magic of this smart self-guiding memory wire instantly transform a professional tie knot around your neck no hands included.

8. A lot of the sneakers has the word air in them, but it has nothing to do with levitating it just means it use pressurized air in a durable flexible membrane to provide light way cushioning to the sneakers. But just imagine if you could literally walk on air. Black budget has developed and designed the vendiair walkers. These awesome shoes will allow you to walk on air up to fifteen feet. The vendiair has magnetic levitation power. These vendiair walkers comes with an electrical air matt together so you can accomplish walking on air.

9. When we think about walking on water theirs only three people according to the bible has done this. Jesus, Peter and Pedro. Whether you believe this or not it still would be considered a miracle phenomenon. Black budget has fiddled with the idea and invented the messiah skates. These water skates will not only let you walk on water, but you will be able to run up to 20 miles per hour. The Messiah skates has jet propulsion under both skates that are one size fit all with the adjusting settings. There are also three rotating fan blades on each side with an electrical impeller pump roller attached to the Messiah skates. Once you turn them on, you'll be able to walk on

lakes, oceans and pools. This will be another way of having water fun.

10. A tow truck is used to pick up damaged or disabled vehicles. If your car breaks down calling a tow truck can be costly if it's not added to your insurance policy. Also, waiting for a tow can be a time-consuming experience a lot of people don't like. Black budget has developed the self-guiding portable towing axle wheels called smart dolly pull. This 300-pound artificial intelligence robot has a GPS system connected to the satellite. This works by activating your portable tow smart dolly pull that will automatically elevate your front wheels lock them both in place. Then put your location to where you're going, and the portable tow smart dolly pull will give you 10 miles charge to your destination.

Chapter 8: Schools and educational leap

1. When we think about calculators it used for making mathematical calculations on a small electronic device. This simplifies the equation while giving you the answer. Black budget took the calculator invention to a whole new level with the new compound fuse together creation design called the totalizer. This one of kind calculator smart pen not only writes down mathematical equations it also displays the correct answer every single time. The Totalizer smart pen has a built-in smart chip that combines with an electric circuit logic gate.

2. School teachers has a very important job. When it comes to educating students and right now in America there's a major shortage of teachers. Due to small percentage of certified school teachers willing to work the students suffer. One day this will be a thing of the past. Black budget has developed and manifested an artificial intelligence classroom cyborg teacher called android tutor. This human like cyborg will be able to administrate lessons also interact with kids and answer questions. This will eliminate real human teachers in the classroom one day.

3. Kids who attend school carry bookbags to hold pens, pencils, books and homework assignments. You never imagine a bookbag being used for emergency safety defense for children. When kids get abducted by strangers on the way to school or after can be a real scary situation for children who are defenseless to adult predators. Black budget has created and developed an emergency safety bookbag for kids called the protection gadget sac. This one of kind protection gadget sac is equipped with an automatic pepper spray button and an artificial intelligent mini robot hand glove that can extend with punching power of 1,600 joules. Also, this incredible bookbag comes with a stranger danger alarm that sounds up to 129 decimals. This protection gadget sac will be a great deterrent for children who could be abducted or kidnapped by strangers.

4. When we think about vending machines being placed in schools' snacks and drinks come to mind first. Vending machines can be used for many different items to store and sell. The percentage of high school students having sex is at all time high and unwanted pregnancy and stds is a problem for early sexually active students. Black budget has designed and created a vending machine called abstinence. This vending machine will come with preloaded condoms, pregnancy tests and stds doctor referrals. This abstinence vending machine will eliminate high school students from being scared to

tell their parents or school medical counselors about their sexual activities. This discrete vending machine will assist students in their time of need when your faced with sexual activities issues.

5. The first day of school for children can be a very scary one because of the separation from parents going into an unknown atmosphere of strangers they don't know. This new experience can be very overwhelming for a child's first day of school. Black budget has developed the I'm still here mommy and daddy bracelet called by your side. This plastic light weight bracelet has a prerecorded button so you can say pleasantry memos to keep your child calm and relax as if you never left their side. This by your side bracelet will be a game changer.

6. Being in class during a teacher's lesson sometimes can be boring and keeping students' attention can be a challenge. The students will do anything to get out of the classroom such as saying they have to use the bathroom is a good enough reason to get out the classroom. This will be a thing of the past. Black budget has invented the body heat temperature monitor chair called BRB seats. These bathroom break seats will turn green if you really have to go to the restroom and alerts the teacher of the situation. This will eliminate students just trying to bale out of class with bathroom break excuses.

7. A chalk board is a hard surface that is used for writing and drawings a lot of the times. Teachers also use it to administrate their lessons. This requires a teacher standing at the chalk board. This could be a thing of the past. Black budget has designed the digital hand free chalk board called stick transfer. This hands-free chalk board comes with the laser writing pad. Every time you write on the hands-free writing pad it will show up on the hands-free chalk board. Eliminating the teachers getting up every time to administrate lessons on the chalk board.

8. The future of school buses and drivers will be obsolete one day. There are a lot of human errors sometimes when dealing with children such as not being dropped off at the right school or leaving students on the school bus. Black budget has created a student drone carrier called fly pass. This artificial intelligence drone has ten propellers with a carry gage on it. This works by calling the customize code given by school you attend. When the fly pass drone arrives to your home all you have to do is select the school you go to and type in where the address is, and the carrier will take you to school. This is the brink of a whole new wave.

9. High School lockers are home based for students and a place to socialize with friends and store your belongings in a secure place. The worst thing that can happen if someone steals your locker combination number and has access to your

belongings. Black budget designed a new android smart locker called ID me. This one of kind smart locker has an identification thumb recognition button also a gun and drug detection monitor fuse that can alert security guards on post. This smart locker also has alarm speaker that will let you know your late for class. This futuristic locker will be a game changer for schools.

10. Delany cards are used in school so the teacher can keep track of student's attendance, grades and determine seating arrangements in classroom. Black budget has created the academy check card. This academy check card will digitally have your First, last name and home contact information each student will be assigned a card that will be connected to the main frame academy check box. If a student doesn't show up to class a alarm will be sent next to their name to alert the teacher of their absence. If a student is absent more than three times the parent will be notified with the academy card. You can earn points for good attendance and will be awarded with gifts and school credit.

Chapter 9: Back II the Future

1. A cellphone case is used to protect the electronic parts inside the cellphone if it drops. Most of the time the cellphone case protects the cellphone from damaging the screen. Cellphone cases comes in many different styles and colors. There's a whole new way to use and enjoy your cellphone case. Black budget has designed and created a live ant farm cellphone case called colony. This one-of-a-kind cellphone case has a fused in real sand with an open and closed latch slide handle. All you have to do is find live ants and put them in this ant farm cellphone case protector and watch how they communicate, demonstrate order and have a working operational system in place. This could way to pass time by just looking at the cellphone case protector. Fire ants would be the best choice to use for this ant farm.

2. Snow is beautiful to look at and play in can also be a nuisance at the same time when you have to shovel your driveway or your car out a parking spot. There are a lot of products on the market to tackle this task but sometimes not as good as it should work.

Black budget has developed, and invented rock salt sodium chloride plastic cutaway snow mats called melt away. This incredible melt away mat comes in a 100 ft roll. You can cut it and put it down where snow is heavy. The melt away snow mats will dissolve snow instantly.

3. A limbo stick is a pole that is gradually lowered from chest level to see how low you can go without hitting the pole. The limbo pole is manually used by a person every time you clear it and emerged to the other side. Black budget created a brand-new version limbo pole set called auto stick flex. This incredible updated version is self automatically motion censored with robotic technology fused inside. The auto stick flex comes with an electrical chest to knee mechanism device that receives an activation signal every time you clear over the auto stick flex. It will keep going lower and lower on its own. No assistance will be needed for this new limbo stick version.

4. When we think about balloons, we associate that with birthdays and parties. Balloons are fun to look at and even self-blowing them up can be fun as well. Some balloons require helium in them imagine a single a balloon that can lift a person off the ground. Black

budget has developed and invented a balloon that can lift a single person off the ground up to 8 feet in the air called airchute. Airchute is a single balloon with a gravity and drag. This 15-point diameter air chute can lift a single person up to 300-pound capacity human weight. The air chute comes with a hand glove attached to the balloon. This is an incredible new way to play with balloons.

5. A chia pet is a clay animal figurine cover with sprouting seeds. Chia seeds are moisturized and patted into the animal and grows when it absorbs water. It takes only weeks to see results. Black budget took the same idea design of the chia pet and developed a new creation called Bob Marley hybrid tree. You can actually grow smoking cannabis with this Bob Marley hybrid tree. All you need to do is administrate chia herb seeds on the Bob Marley hybrid tree. Water the tree and keep it under a grow lamp and within three weeks you will have ready to smoke cannabis. You can grow up to 12 grams of marijuana with this Bob Marley hybrid tree.

6. Most people who perish in a fire die from a lack of oxygen due to the toxics gases making it hard to breathe. This is the main

cause of death before the body is consumed by the fire. Black budget has invented the safety fire helmet called bubble head. This strong plastic one time use can buy you some time. It works by putting on the bubble head and pulling activation string automatically releasing oxygen through the ventilation flow cord attached to the helmet keeping you alive. The bubble head is truly a game changer that will change lives.

7. When we think about having personal protection, we associate that with a gun. Which is the most effective deadly weapon that can give us a peace of mind. If your ever faced with any type of danger, it's the ultimate choice for protection. Black budget has developed and invented an alternative protection weapon called the defense snappers. Defense snappers are tiny 50 pellet concoction ball that's thrown on the ground that shoots and outward explosion to deter you out of danger. It can't kill muggers or burglar but its excruciating pain if your hit with this defense snapper.

8. Eating healthy is a personal choice and becoming a vegetarian is definitely a lifestyle change. Knowing what and not to eat as a vegan is very important. Trying to prepare your own vegan meals that taste

good can be difficult for people that's never done it. Of course, they have cookbooks that can assist you on making vegan food, but black budget has taken all the guess work out with the invention of the talking meatless vegan maker. This incredible state of the art technology device will allow you to prepare meatless good tasting food with one button. If you want meatless chicken, pizza, burger and etc. all you have to do is follow the talking recipe ingredients settings. Set the correct timer as directed and watch how this device will make amazing vegan meals. Eating healthy will taste so good.

9. A portable photo booth allows you to snap pictures of you and your bestie, relative or boyfriend. It's a fun way to capture them priceless moments we experience in life. Black budget has invented a kiosk machine that will allow you to shoot music videos called capture all in one. This will ultimately eliminate the need for actual videography if you choose to. This incredible music video shoot booth is equipped with a digital green screen with over 10,000 video background scenes you can choose from and five cameras inside. All you have to do is choose one or more of the scenes props you want to use and link the booth to your

electric device through Bluetooth to upload your music. Then create your music video and when your done the kiosk will edit, finalize and send video to your email.

10. Having wild animals as a pet can be dangerous or fatal. They do not adjust well in captivity and are unpredictable. They also eat more than the average pet which can make it hard to care for them. Looking at exotic animals and being able to touch them or play with them are two different things. The average person will never know how it would be to have a real-life wild animal as a pet. Black budget has invented and created Mach 5 animal predator robot toys called RLP. These realistic predator toys are controlled robots you don't have to worry about cleaning after or harming you. Lions, Tigers and bears can now be your pet without any danger.

Chapter 10: Exploration

1. Space travel is a real thing that our shadow government has already been doing. They have a missile space craft called the majesty 2.0. It has ventured out of space to different star systems. This one of kind space craft is occupied with artificial humanoids who send back real data to black budget operational team support.
2. Life on the surface of earth is just one existence. You would be surprised to know that there is more advance monetary living system underground. The only way you can get down there is if your very wealthy and picked out the lottery pool. Black budget has designed a state-of-the-art navigator go cart that will take you places underground that doesn't exist on the surface such as a pond filled with beautiful male and female mermaids that you can swim with. Also, theirs a nudity mall underground called birth suits and a skinny-dipping club for the young and sexy called Sexy dipping.
3. Prisons in America justice systems hold more than 2.3 million inmates. There are prisons on earth that are exclusive to the general public their locations are unknown. Black budget developed and designed a state-of-the-art prison on mars that hold humans and different life forms. If your ever

end up in this prison on another planet, you'll never see earth again.
4. When we think about how the extremely wealthy one percent vacation. We think about expensive hotels such as the Ritz, Saesor Place or the Plaza Hotel. These are just a few choices on how the extreme wealthy vacation. You be surprised to know black budget has architect infrastructure design a vacation resort on a planet called Kepture. This planet is similar to earth. The only way you can go to this resort is to be a part of an upper high-class society.
5. Dinosaurs ruled the planet earth over 65 million years ago. They became extinct due to an asteroid that penetrated the surface of the earth. A lot of them died due to lack of resources. A lot of fossil bones have been buried all around the world. Black budget has collected a lot of different bones. They we were able to rejuvenate these prehistoric creatures on a planet called Apex where they thrived and still exist today. This planet is also a vacation destination for very powerful wealthy people. The only way you can get to this planet you must be a part of an upper-class society.
6. It may be hard to believe that other life forms actually do exist and habitat other planets, but they do. Would you believe there's another planet called Swoof where

little blue beans exist? There was a cartoon out that depicted these blue beans. In all actuality they do reside on the planet called Swoof.
7. Mars is the fourth planet from the sun. Would you believe Mars has a human life underground base where only powerful and extremely wealthy people live. They have stables where horses race and also an exotic zoo. In the exotic zoo they cross breed animals with other animals and humans, so they look very different from regular zoos on earth. It's a little bit of freak show. They also have a sex slave cloning lab. So, you can make your favorite celebrity or anyone your attracted too. The rules on earth doesn't apply there.
8. There are over 400 billion planets that can sustains human life form. This one amazing planet called Gliese's that produces natural sunshine rays. It makes thousands of rainbows at one time. Because this planet has high pressure chemical vapors when it starts raining it makes diamond flakes. This is a very pretty planet with colors that do not exist on earth.
9. Deep space abyss travel has already been going on for a very long time. This is navigated by genetically modified cyborg machines that use stargates to travel to other planets such as planet Zuplous. This planet

is occupied by high vibrational giants who communicate using telepathy.
10. The universe is masked with infinite space. The Milky way galaxy has over hundred-billion-star systems that support different life forms. There is a planet called Arista Mermayde where humanoid mermaids' habitat. These incredible looking human facial features with fish bodies parades the seas in the thousands. On this planet they communicate with people with bubble talk because that's their fluent language.

Closure

If you took the time to purchase my book, then you are reading this right now. I just want you know that I really appreciate you so much. Thank you. When I started my journey to write this book, I envisioned a future with supreme technology that could possibly already exist.

I know a lot of times when products are introduced to the general public such as cell phones or game consoles. They have already manifested a greater version than the previous one. So, the technology can be far greater than what we can imagine it just has not been presented to the masses. LOL my friendly readers sorry to inform you that this book I wrote is a fabrication of ideas and thoughts straight from my brain. None of this technology that I mentioned in this book actually materialized yet to my knowledge but just keep in mind readers nothing is impossible.

CPSIA information can be obtained
at www.ICGtesting.com
Printed in the USA
BVHW012204021222
653297BV00009B/773